APPRECIATION

To my friend Justin Titima,
Thank you for making this book possible.
May our friendship continue to flourish from generation to generation, up to eternity.
God bless you!

So he said to me, "This is the word of the Lord to Zerubbabel:
"Not by might nor by power, but by my Spirit"
Says the Lord.
(Zechariah 4:6)

All glory to You Lord!

THE BASICS

Before we sail and launch deep into this topic, it is imperative that I mention a few things which are of great importance.

Passion

You need to have a passion for mathematics. This is what will fuel the fire in you to start and continue applying the strategies in this book. It is what will cause you to substitute other things in your life so that you will find time to put into practice what we are laying down here. Passion is what will keep you going even when other people have fallen by the wayside. And in the end, the results will be as clear as crystal.

Without passion you will never make it.

It is passion that awakens a desire in the heart of a young boy living in the deep rural jungles to walk for days or even weeks, going to the urban land because he has heard that there are opportunities there that can change his life forever. His peers in the county-side may still be waiting for that perfect chance; a line of least resistance which may never come.

Change Your Mindset

For you to be good at maths, you will have to change your mindset concerning the subject.

If you are the kind of person who has always had an attitude that "Maths is a difficult subject, and only clever people can understand it" then you will need to change that kind of attitude.

It is the same thing that happens when a person wants to lose weight. They must change their attitude about exercising their bodies and the amount of food they consume; that is diet.

Know the fundamentals

It is imperative that if you are going to use this strategy right you know the fundamentals of mathematics. These are the foundational truths that govern maths and where the whole of the subject is grounded.

The basics are subtraction, addition, multiplication, and division, the subjects that are commonly taught at the lower grade level in school.

Write, speak, hear

Do you know that in the Jewish traditions, whenever the king ascended the throne he would be given a copy of the commandments of God?

Then he would be instructed to rewrite the same laws on a different paper in his handwriting.

Try to imagine; This king had clerks, prefects, and servants. He had governors and every other person who could do this simple task for him. Writing this thing was simple. It would just take a couple of minutes and it was done.

But the ancient savants knew something important. They knew that there was a connection between the mind and the heart whenever someone would write down something in their handwriting which would make them not only remember but also do it.

So I suggest that every exercise you are going to do in this book needs to be written down. You will need lots of exercise books or foolscaps,[Which ever you desire] to write and rewrite endlessly.

On top of that, you have to read loudly the formulas and exercises you are going to do in this book. This will enable you to hear yourself which works well with other people.

Repeat,repeat,repeat.

Everything you are going to learn in this strategy has to be repeated continuously.

Do not try to cram anything. Just repeat it.

Repetition is what makes human beings master a lot of things. For example; if you are learning to play a new musical instrument or are learning to speak a new language, the key is repetition.

As you continue to repeat whatever you are learning and as the ideas begin to pop up in your mind like popcorn, then you know that you are on course to cracking the code that will enable you to know all the mathematics and lay it bare at your feet!

Repeat; do not cram!

What you want to know

You will need to spell out what you want to know. Do you want to know the whole mathematics syllabus or do you just want to know a portion of the subject?

MATHEMATICS SYLLABUS

Someone did say that to succeed in anything,

[1]You need to know what you want.

[2]You need to have the right strategy to get you there.

[3]You need to apply that strategy.

In this chapter, we get you running into the strategy that will get you into the system which will make you well-versed in mathematics.

The first thing we need to do is to know the whole scope of the topics you want to know.

We call this "Mathematic syllabus."

I have named down the topics in the whole caboodle of the syllabus as taught in grade schools and high schools. This primarily covers almost all topics taught in mathematics.

[1]Great common divisor.

[2]Divisibility tests.

[3]Fractions.

[4]Numbers

[5]Symbols

[6]Squares and square roots.

[7]Decimals

As you can see, we have been able to write down just over fifty topics in this mathematics syllabus. These, I believe are almost the maximum topics taught in math. In general, each topic has a minimum of five subtopics. For example; under the topic of Square and square roots. The subtitles are:

(1)Squares by multiplications.

(2)Squares from tables.

(3)Square roots by factorization.

(4)Square roots from tables.

When we have the full information as we have underlined in this chapter, it becomes easier to chart a way forward. In the real sense, we have a road map to get us to our destination.

What I mean is that people think that the subject of mathematics is infinite; never-ending. That is, the topics it covers just go on and on.

At first, I used to think in that manner. I used to believe erroneously that the topics in maths are more than three hundred or even more; like a thousand. And I was intimidated by it. That thought made me think that maybe that was the reason they kept us in grade school and high school for so long.

But when I realized that they were just above fifty topics to be covered in both grade and high school, I was elated. Psychologically it made me positive to know that maybe there must be a formula that I could use to crack those more than fifty topics and master them.

Here is the formula.

TOOLS

Before we go deep into this topic, there are things that you need to have to enable you to succeed in your endeavors. These are the tools.

Before anyone undertakes a journey, they must see that they have carried with them everything they need. Failure to do that will mean a doomed voyage.

Mathematical set.

This set contains all the instruments you will need to do your maths. You will have to do everything practically as it is required in maths. That is why you will need pens, rulers, a compass, a protractor, a square, rubber, a logarithm book, e. t. c.

Two hardcover books.

You will need two hardcover books. These act as your reservoir to store all the information you have collected about mathematics. We will talk more about these important books in a later chapter.

Foolscaps/exercise books.

In this exercise, you are going to do a lot of writing. You will need plenty of foolscaps or exercise books, whichever you prefer. Because you are going to write and rewrite, you will need a clean, clear space where you can write clearly. Remember that your mind is able to pick whatever you write down and bring it to your recollection. Whatever you write has to be clear to enable your mind to store up well the materials you are feeding it.

The key.

In order to unlock anything that is locked up, you need the key. And so is maths. In this strategy we are setting up, you must have a key. The key is so simple because you do not need to set up your own. It is already there.

As in any Self Teaching book, you do not have to make your own key. You use traditionally what has always been there.

What you do is that you use the God-given gift in you to create a strategy by utilizing what traditionally has always been there.

For example: Right now I am learning to play the guitar using a Self-teaching book given to me by a wonderful lady, a sister Jessica from New Song Church, Tennessee. The book is easy to read and the information they give in that book is fantastic. I am sure that as I continue to do what the authors are saying, I will know to play the guitar without any man's help just by reading it from the book.

This is what I mean by using what has traditionally been there. The authors of this great book have not formed their own cords, scales, musical notes, staff, and dynamics, just to mention but a few. But what they have done is that they have taken all the musical stuff that has always been there,(the ones we have named and more.) and have just added a little thing of their own which makes their book different from the other music books. And this little but powerful thing is the God-given ability to use what has always been there to create a strategy out of it.

This then enables you as a teacher to chart a path for the student to follow so that he can be able to master easily what has always been out of his grasp.

That is exactly what I am doing in this book. I am taking the mathematical materials that have traditionally been available and am using the God-given gift that He has imparted to me to create a strategy to show you, the learner an easier way to master mathematics.

These traditional truths that have always been there are contained in the current mathematical textbooks that contain all the fifty-five topics we have mentioned in the previous chapter. The only thing that these books miss is the strategy to help you master all these topics at a go.

I have the master key!

What you need to do is to collect as many textbooks as you can that cover the topics that we have mentioned. It is your passion that will separate you from the rest. Buy the books if you have to. You will never regret it.

The most important thing is that the material you collect must be genuine. If you collect what is fake, it will work against your development because it becomes difficult to unlearn it later. It is like learning to play music with fake cords only later to try and rectify it. It becomes difficult; almost impossible.

STRATEGY (A)

The old Zulu warrior had two lethal weapons; the spear and the assegai. The spear was for long range, to fight the enemy who was afar.

With the spear, the warrior could hurl it at the enemy who looked fierce or even who was trying to run away. The only problem with the spear is that the warrior had no guarantee that he could hit the enemy, and also he could just hit a single enemy with it. He must have had a great deal of practice to make sure that every time he threw his spear he would hit his target.

Whatever we are going to discuss in this chapter is the spear.

But with the assegai, it was different. First, the warrior did not have to lose it by throwing it away. He always had it in his hand. The assegai was like a spear, in that it had a head like that of a spear. The difference is that it was heavier and had a short handle.

This weapon was lethal. With it, the warrior could kill many foes. If the enemy came close by, like some feet away, he could rip him with it or smash their skull or chest with one blow and take it back.

In the next chapter, we are going to look at the assegai.

We are not saying that this chapter is not important.No.Just the way the warrior knew that both weapons were important. He had the realization that some enemies would dare not come close by. He was conscious that if they escaped, they would disappear forever. He had to shoot them down.

Now, let us get down to the subject.

Take one of the hardcover books. In the last chapter under the subtitle 'Key', we said that you had to find the right Textbooks that are going to give you all the information about the fifty-five topics we have mentioned.

For example; starting from the first topic that we mentioned 'Greatest Common Divisor' mentioned in the mathematical syllabus, write down the Topic, the Formula for that Topic, and an example for it in one of the hardcover books.

Make sure that the material you write in this book is genuine or from credible sources.

If there are any other subtopics to be covered under that topic, write everything down. Mostly, the Topic has a minimum of five subtopics.

Arrange everything systematically as we have lined them down, from the first one to the fifty-fifth one.

For example; if I choose randomly from the Topics we have written down and choose 'Squares and Square Roots; I will write it like this:

TOPIC:

SQUARES AND SQUARE ROOTS.

FORMULA:

A number multiplied by itself.

$X2$.

EXAMPLE:

6 X 6=36

Starting from the first topic we have written down; that is 'Greatest Common Divisor' to 'Matrices and Transformation', you write them down like that.

That is why I suggest you have a big book because that way I believe it will be able to carry all the contents you need to put down.

The reason I like hardcover books is twofold. First, by its hardcover, it is protected and can stay for a very long time. Second; they are the few books that can be gotten in A4 and have many pages. (Like 500 pages). You need all your materials under "one roof", that is, in one book.

You can draw your materials from every channel you get provided that it fulfills what you want and that it is authentic.

After finishing writing all the Topics, Formulas, Examples, and any other things you need in every topic, you will have a nicely laid out map to chart you into what you need to learn.

Now, this book, not the Textbooks, will become your reservoir for any mathematical problem or study you want to solve. Anything you need to look up concerning mathematics check in this book, not the textbooks.

How are you going to use this book, or how important is this book?

You are going to use this first book to learn, relearn and practice what you have learned.

This is how you are going to go about it.

Every single day, you are going to take one topic. Since a single topic has a minimum of five subtopics, you divide that single topic by two. That means that you are going to take two days to study one topic.

On the first day, you take half of the topic's subtopics and just check them out. If there are five subtopics, you can take two or three of them. If there are eight, you can take four.

Almost all the topics begin with a formula and then an example. If there are other things, they follow later.

You look at them and if possible read them out aloud. Do not try to overwork your mind and of course, do not try to cram anything. When you finish, close your book.

The next day, before you begin tackling the other remaining half. What you do is that you revise what you had done the previous day.

You reread what you had done the previous day. This time around, after you have gone through the materials, close your book and try to write down what you have just learned in an exercise book or fools-caps. If you have to write a formula and an example, write them down without looking at them from the book; just out of memory. Then after you have finished writing everything down, correct yourself to see whether you have done the right thing. Do not worry if you have not done it well. Repetition will take care of the problem.

After you have finished, now turn to the remaining half of the topic that remained. Do the same thing you have done with the first half. Go through it; look at the formulas and the examples, and speak them out aloud. You have finished. Do not do the correction of this second half. That is for the next day.

I find it easier to learn when I see what I am learning, that is writing down everything. You will need to do these three things: write it down, speak it out so that you may hear it, and then see it from your handwriting.

The whole exercise can take about fifteen minutes.

To go through this whole book in this way, you will need a total of about 110 days; that is the number of topics times two,(because you have divided the topics in half). That comes to say 55 x 2=110 days.

That will come to about three and a half months. That means that every year you will be able to do the exercise three to four times if you do it daily; which is good but not appropriate.

When you learn something once after every three and half months, it is not easy for you to remember it easily. You will need to introduce another system that will enable you to do the whole exercise in a shorter time; maybe three weeks or even less.

When you do it in three weeks or less, it becomes easier for your mind to recollect easily what you have learned.

This takes us to the second strategy; the assegai.

STRATEDY (B)

If you are going to be able to execute correctly what we are saying in this book, then, learning maths is going to be easier and more enjoyable for you.

And what we are going to talk about in this chapter is easy because you have to create something that fits you personally.

This is what I mean: Out of the topics we have been able to write down, you will need to concoct an acronym(s) that will be able to cover all fifty-five topics.

For example, The topics we have named are as follows:

(1)Greatest common divisor.

(2)Divisibility test

(3)Fraction. E.t.c

Let us say that if you take the first topic we have mentioned; which is 'Greatest common divisor', and because it begins with the letter 'G', you say that 'G' stands for 'Greatest common divisor', then 'D' stands for 'Divisibility test', and 'F' stands for 'Fraction' and so on and so forth.

So, if I write G, D, F, what I actually means is 'Greatest common divisor, Divisibility test, Fraction.

But because the topics are so many, we know that that is not applicable. Furthermore, there are other topics that start with the letter 'G' or 'D', or 'F'.

The goal here is that you become creative and create something personal that will be able to cover all the topics for you. You do not need to arrange the topics systematically the way we have written them down. You can even write the first one last and the last one first. You may even need to introduce other letters or even numbers to accomplish your mission.

Let us get deep into it.

We are going to write down every topic and then beside it write the initial or first letter that represents it.

For example: If we write the topic 'Pythagoras Theorem' then beside it we write the letter 'P' to show that it represents that topic.

So, we start writing in the same order we have written them; from the first one to the fifty-fifth one.

(1)Greatest Common Divisor. (G)

(2)Divisibility Test. (D)

(3)Fractions. (F)

(4)Numbers. (N)

(5)Symbols. (S)

(6)Squares and Square Roots. (S)

(7)Decimals. (D)

(8)Factors. (F)

(9)Integers. (I)

(10)Least common multiples. (L)

(11)Linear motion. (L)

(12)Compound proportions and Rates of work. (C)

(13)Approximation and Roots. (A)

(14)Reciprocals. (R)

(15)Cubes and Cube Roots

(16)Area. (A)

(17)Measurement. (M)

(18)Rates, Ratio, Percentages, and Proportion. (R)

(19)Further Logarithms. (F)

(20)Indices and Logarithms. (I)

(21)Area of Quadrilaterals and Other Polygons. (A)

(22)Time. (T)

(23)Volume of Solids. (V)

(24)(Volume of Capacity.(V)

(25)Pythagoras Theorem. (P)

(26)Equation and Straight Lines. (E)

(27)Algebra. (A)

(28)Mass, Density, and Weight. (M)

(29)Surface Area of Solids. (S)

(30)Area of a Part of a Circle. (A)

(31)Sequence and Series. (S)

(32)Surds. (S)

(33)Geometry. (G)

(34)Quadratic Expressions and Equations. (E)

(35)Linear Equation. (L)

(36)Scale Drawing. (S)

(37)Loci. (L)

(38)Linear Inequalities. (L)

(39)Formula and Variations. (F)

(40)Binomial Expressions.(B)

(41)Graphs. (G)

(42)Circles. (C)

(43)Commercial Arithmetic. (C)

(44) Three-Dimensional Geometry. (T)

(45)Common Solid. (C)

(46)Vectors. (V)

(47)Statistics and Probability. (S)

(48)Angle Properties of a Circle. (A)

(49)Trigonometry. (T)

(50)Graphical methods. (G)

(51)Integration. (I)

(52)Elementary Calculus. (E)

(53)Navigation. (N)

(54)Area Approximation. (A)

(55)Matrices and Transformation (M)

Above are all the topics and the initials or first letters that represent them.

Now, we are going to write again the first letters alone of all the fifty-five topics we have mentioned. For example: "N" which we know may stand for "Navigation" or 'Numbers'.

Here we go:

(1)G	(13)A	(25)P	(37)L	(49)T
(2)D	(14)R	(26)E	(38)L	(50)G
(3)F	(15)C	(27)A	(39)F	(51)I
(4)N	(16)A	(28)M	(40)B	(52)E
(5)S	(17)M	(29)S	(41)G	(53)N
(6)S	(18)R	(30)A	(42)C	(54)A
(7)D	(19)F	(31)S	(43)C	(55)M
(8)F	(20)I	(32)S	(44)T	
(9)I	(21)A	(33)G	(45)C	
(10)L	(22)T	(34)Q	(46)V	
(11)L	(23)V	(35)L	(47)S	
(12)C	(24)V	(36)S	(48)A	

The next thing we are going to do is that we are to collect the letters with the same initial together. For example, We are going to take all the letters with "F" and place them together. These letters are Factors, Figures, Further Logarithms , Formula and Variations ; which are four.

This is how we are going to write it since there are a total of four topics that begins with the letter "F" then we write it this way:(4) F. And we continue that way, writing every letter that represents every topic until we have finished.

Here is how we do it:

(7)S,(2)N,(4)F,(2)D,(4)G,(4)L,(3)I,(5)C,(2)R,

(7)A,(3)V,(2)E,(1)Q,(3)M,(1)B,(3)T,(1)P

Above, we have gathered together all the initials that represents all the topics we want to study. All that start with the same initials e. g "R" are two. These two are "Reciprocals" and "Rates" e. t. c.

If you add up all the numbers before the initials ,they add up to fifty five, the full number of the total topics.

What we are aiming at is create an acronym(s) that is short enough and that will enable us when we mention it or them will be able to contain all the fifty five topics.

The simplified initials we have now that is if we remove the numbers before the initials add up to seventeen.

These are the same letters we have listed above but this time without the numbers:

S, N, F, D, G, L, I, C, R,

A, V, E, Q, M, B, T, P,

From these seventeen letters, we want to create something personal, an acronym(s),that when we mention will be able to cover all the mathematics topics.

These are the four acronyms I have been able to create out off the seventeen letters.

(1)FLING.

(2)VEST.

(3)C R A M B P D .

(4)Q

As you can see, the acronyms do not necessarily need to make any sense, that is ,in the way they are spelled. Take for example, the one C R A M B P D. It does not make sense ,but the reason I choose it is because it sounds like a familiar word C R A M P E D, and it is also easy to memories. The aim is to achieve your goal.

One thing we must remember is that these acronyms carry all the topics with them.

For example: When you write FLING; it stands for

(4)F, (4)L, (3)I, (2)N, (4)G.

This means that in the four representing "F" the number four represent all the symbols that begin with "F"; Fractions, Further Logarithms, Formula and Variations, and factors.

The same applies with the (4)L, with the (4) representing the four topics beginning with L; which are :L, C, D, Linear Motion, Linear Equations, Linear Inequalities.

The list goes on in the same manner.

That means that in order to make it easier for us, when we read the acronym FLING, what that actually needs to come into our mind is this:

(4)F, (4)L, (3)I, (2)N, (4)G.

When we read the acronym VEST, what that needs to come into our mind is

(3)V, (2)E, (7)S, (3)T.

When we read the acronym C R A M B P D, what that needs to come into our minds is :

(5)C, (2)R, (7)A, (3)M, (1)B, (1)P, (2) D.

And of course when we read the acronym Q, what that needs to come into our mind is this:

(1)Q

We need to commit to memory these four acronyms in the order that we have lined them with the letters and the numbers with them.

When we do so, we make it even easier for us to know and say by heart all the topics we need to cover because we have the words plus the numbers.

STRATEGY (C)

Now that we have concocted acronyms that will be able to personalize all the topics we need to cover, we go to the next important thing.

But just as a reminder, our four acronyms are these:

(1)FLING

(2)VEST

(3)C R A M B P D

(4)Q

We have said you need to memorize them like this:

(1) (4) F, (4) L, (3) I, (2)N, (4)G

(2) (3)V, (2) E, (7) S, (3) T

(3) (5) C, (2)R, (7)A, (3)M, (1)B, (1)P, (2)D

(40(1) Q

With these, you are sure to cover all the mathematics that you want to know.

In this chapter, we are going to cover the importance of the second Hardcover, A4, 500-page book.

In this second book, you are going to rewrite all the topics you wrote down in the first book. Only this time round, you are not writing them systematically the way you had arranged them in the first book.

No; you are going to rewrite everything in the way they appear in the acronyms we have created.

The first acronym is FLING. We have said that in your mind, this should appear like this:

(4) F, (4) L, (3) I, (2) N, (4) G

That means that the first topic in your book will be one of the four beginning with the initials "F".

It could be Fractions, Factors, e .t. c. You write down the Topic, Formula, Example, and any other thing needed in that topic.

What you need to do is that you need to decide that of the similar letter representing topics in a group, which one will be first and which one will be last

For example; In the three letters that represent "V" in the acronym VEST; That "Vector", "Volume of Solids", "Volume and capacity", you will have to decide that the number one in your mind and the book as it appears should be "Vectors" then followed by "Volume of Solids" and then "Volume and Capacity".

Do not put this one first and then the next time you put it last. No. You need to organize it well for your mind so that it can compile it nicely as in a computer.

Writing all this work down will take time. It might take weeks or even months. That is why we mentioned the importance of passion.

After you have written down everything, it becomes easier for you to not only memorize using the acronyms but to also do your corrections. When you mention the acronym FLING to the last one Q, everything just flows out. All the ones with "F". The Topics, Formula, and examples e. t .c.

This second book becomes your companion to help you practice and commit to memory of all the topics in the shortest time possible.

This is how to use this book. Without opening the book, you can say aloud your four acronyms. As you do that, you write them down in an exercise book or foolscap somewhere.

If you have forgotten any of the topics or their formulas or examples, do not copy them from the book. But leave a space in the exercise book which will be enough for you to do your revision.

After you have finished writing all the acronyms, now it is time for you to do your corrections.

Doing your corrections becomes easier because you just flow with everything you have written down in this second book.

Your goal is to write down all that appears in the acronyms without checking. Write them down and to do the corrections should not exceed three weeks.

When you finish everything, you start again. Every time you do corrections and find that you have left a space for something you did not know or had forgotten, copy that thing down exactly as it appears in the book. Copy it down on that space. As you continue with the exercise, repeating it frequently, the empty spaces will begin to be fewer and fewer, and soon, there will not be any space to be filled because you will have mastered all the maths you have in your acronym which you have lined down for yourself.

The key to mastering all the maths in the acronyms is repetition.

Arranging the topics in this book to correspond with the acronyms we have created also gives the mind a clearer picture of how neatly everything follows each other.

It comes a time whenever you mention the acronyms, the pages of the book just begin opening up in our imagination.

You have mastered mathematics. You now have two big books. The first one is for a long-term plan, to take you slowly, to learn and commit everything to memory. It can take three and a half months to four months to go through this book.

The second one is just as important. But this one is for a short period. It is to do the same thing, but in the shortest time possible; three weeks or less.

You need to calculate everything you are going to do in this second book so that it does not exceed three weeks. If it does, it beats the goal of what you want to achieve.

The Zulu warrior carried the assegai whenever he went. When he woke up in the morning, it was by his side. Whenever he did his daily chores, it was in hand. When he walked talked or even had his meal it was close by. And of course, whenever he retired to his bed at night he had it on his bedside; ready to be used.

I want to show you a formula that you can use so that you can carry this strategy; the assegai whenever you go.

RECITATION

Since you are passionate about mathematics, you will have to use all the arsenal at your disposal to keep on improving.

One important artillery in your arsenal is recitation. Recitation will help you keep all the topics with their formulas fresh in your mind continually.

An easier way to do recitation is to use the acronyms we have formed; that is FLING, VEST, C R A M B D, Q.

What you do is that you just recite!

You can call out aloud or in quietness one of the acronyms; say FLING.

You know that each of the letters in the acronym stands for certain topics.

For example: "F" stands for "fractions", and "Figures" and their topics.

Then "L" stands for "Logarithms" and other topics. And so on.

You begin with the "F" since it is the first letter in the acronym. After you call it out, you then mention the topics one by one, mentioning the formula, examples, and any other detail needed. Remember that the "F" has four topics.

When you are done with the letter "F" you then go to the letter "L" and then do the same thing as you have done with the "F". You continue that way until you have finished all the letters in the acronym

FLING.

You then take another acronym, VEST. You repeat with it everything you have done with the first acronym. You continue like that until you have finished all the acronyms.

You can not be surprised that in one day you can be able to do these recitations a couple of times.

In this way, you will be able to improve your knowledge of mathematics even more.

All these depend on one thing: The passion you have for the subject.

The warrior never walked with his assegai like a boring weapon. He juggled it the way the circus man juggled his instrument; throwing them into the air and then catching them again. Passing it from one hand to the other the way the basketball player played with his ball. This is how he did it.

MATHS CARD GAME

To fully utilize this acronym strategy, you can play the maths card game. This game is played the way one plays cards, only this time round, you are practicing maths.

Just like playing cards, this game can be played by two or more people.

What you need is a lot of fools-cap or exercise books.

This is how it is played. First, everyone by memory writes down on a paper all the fifty-five topics of maths. It is easier to do it when you use the acronyms. This exercise alone by itself helps the learner to be able to memorize the topics.

Then, the first person names a topic that he wants his counterpart to cover. Afterward, the other person also names a topic he wants the other person to cover. If there are more than two people, each person in the circle calls a topic they want the next person to him or across to him to cover until each person has a topic.

Then each one goes to work.

To avoid repetitions, whenever anyone asks a question, they tick it out from the list that they have made. The game ends when all the fifty-five questions have been asked. You can start again if you want.

Each person needs to write on the foolscaps the questions they have been asked to cover. If they need to define it, they can. In each topic, they give out the formula, examples, and any other subtopic needed in that area.

After they finish ,the next question is asked, and everyone repeats the same until all the questions are answered.

Each fools-cap has to have that person's name.

The easiest way to do the correction together as a group is to group all the topics with the same title. After you have done that, you return it to the owner only the one foolscap with the topic you want to correct at that time.

All of you as a group can then go through your papers by referring to the hardcover book.